Bella dear, The Engineer

Adapted by Billye Boddie

•••

Bella dear, The Engineer

Books may be purchased by contacting the publisher and author at
www.stemtogreatness.org

Publisher: STEM to Greatness, LLC

ISBN-13: 978-1983490897
ISBN-10: 198349089X

First Edition
Printed in the United States

Dedicated to my late Great–Grandmother Evelyn K. Davis,
a champion for families.

Meet Bella

Bella is a creative and confident little girl who loves to explore. She has a sharp mind and is eager to learn new things.

"Bella dear,"

Bella's mom walks into the room and Bella made a request. "Mommy, I want to use my creativity to build a robot. A robot like the ones we saw at the STEM event."

Bella paid attention at the STEM event.
This was the first time Bella saw a real robot.
She was overwhelmed and intrigued but above all she
developed an interest in robotic engineering.

At the STEM event, Bella met robotics engineer Ms. Dream. "Hi kids, I'm going to teach you about what I do as a robotics engineer." The kids cheered at the announcement.

Bella was bold enough to walk up to Ms. Dream and
shake her hand. Ms. Dream was so impressed she
asked Bella to press the start button to power on the robot.

Bella's mother looks at her and says,
"You are a brave girl, you can do whatever you put your
mind to." That was the beginning of great things to
come for Bella.

Bella built up so much confidence, from the STEM event that she wanted to build a robot at home. Bella learned from Ms. Dream that engineering is a step by step process.

"I want to be a robot engineer.
I can do it. I can build a robot."

ENGINEERING PROCESS

DESIGN

First, you design, then build
and last test your creation to
make sure it works.

The first step is *design*.
Bella uses her creativity to *design* the
robot using paper and crayons.

ENGINEERING PROCESS

BUILD

Now that Bella has finished the *designing*, she moves to the next step to *build* it.

Bella imagines the parts needed to build her robot.

Bella looks around the house to find all of
the pieces to build her robot.
"I am going to have so much fun."

"I am bold, confident and creative".
Bella has the confidence to know she can do it.

In the bathroom she finds empty toilet paper rolls to use
for the arms and legs.

In the kitchen Bella finds a cereal box to use for the
robot's stomach and a mac and cheese
box for the head.

"Bella dear, I have some things for your robot." her mom said handing her phone and tape to Bella. "What do I do with your phone?" Bella's mom tells her to press record and speak in a robotic voice. Then place your voice into the robot head.

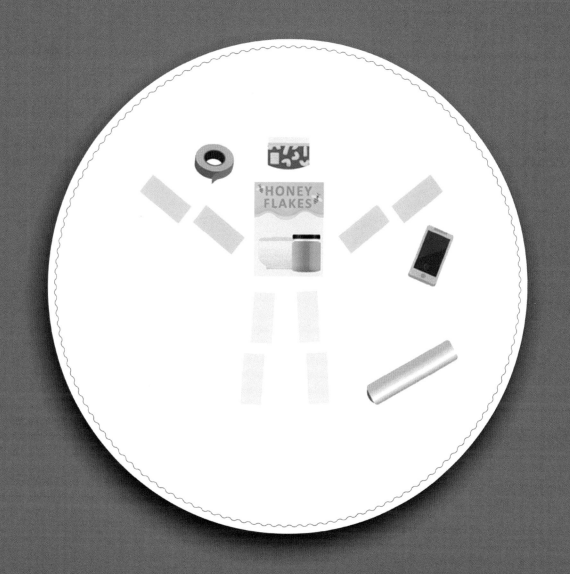

She does so by putting all of the parts together and records a robot voice on the phone. "I am a robot, I will help make the world a better place"

ENGINEERING PROCESS

TEST

Last, Bella will test her robot.

She imagines herself in the lab
testing the robot.

My robot is complete.
"I knew I could do it!"

"Bella dear, is that you?"
"Have you tested your robot yet?"
"Yes Mommy, look"
Bella holds up her creation.
"I'm an engineer Mommy."

Bella used her inquisitiveness and creative skills to find all the pieces of the robot and made them come together to bring to life, what started in her mind.

SCIENCE.TECHNOLOGY. ENGINEERING.MATH

What STEM professional do you want
to be when you grow up?!

BELLA'S TIPS FOR KID ENGINEERS

1. Don't be afraid to ask a lot of questions.

2. Explore the world around you.

3. Use your imagination when you are bored.

4. Read books about scientists, technologies, engineers and mathematicians.

5. Build something from things you find around your house.

6. Imagine something cool in your mind and make it come to life.

LESSONS LEARNED

- When you believe in yourself, anything is possible.
- Bella used her creative skills and confidence first to make the robot then took the engineering design steps to bring it to life.
- Always think outside of the box and follow your dreams.

MEET THE AUTHOR

Billye Boddie is an engineer, author and STEM speaker based out of Atlanta, GA. Her educational books for all ages are based on her own experiences and ideas, and intended to encourage diversity, confidence, and excitement in the STEM fields. Offering youth tech workshops to schools and youth agencies, she uses her books as a companion tool on her quest to help students navigate their future career paths in science, technology, math, and engineering.

Billye's great-grandmother Evelyn K Davis's crusade of the importance of early educational opportunities for children played a major role in influencing her to pursue dreams of encouraging diversity and inclusion in technology careers. Seeing how hard her grandmother worked to make a difference for the better in their community, Billye was inherited with the ability to look for ways that she too could impact the world and people around her in a big way.

Made in the USA
Columbia, SC
13 April 2021